PIONEERING TO AI SYSTEM THROUGH ELECTRICAL ENGINEERING AN EMPIRICAL STUDY OF THE PETER CHEW RULE FOR OVERCOMING ERROR IN CHAT GPT

PETER CHEW

PCET VENTURES (003368687-P)

Email:peterchew999@hotmail.my

© Peter Chew 2023

Cover Design : Peter Chew

Cover Image: Freepik Premium

No part of this book may be reproduced in any form or by any electronic or mechanical means, including information storage and retrieval systems, without written permission from the author.

All rights reserved

Mathematician, Inventor and Biochemist Peter Chew

Peter Chew is Mathematician, Inventor and Biochemist. Global issue analyst, Reviewer for Europe Publisher, Engineering Mathematics Lecturer and President of Research and Development Secondary School (IND) for Kedah State Association [2015-18].

Peter Chew received the Certificate of appreciation from Malaysian Health Minister Datuk Seri Dr. Adam Baba(2021), PSB Singapore. National QC Convention STAR AWARD (2 STAR), 2019 Outstanding Analyst Award from IMRF (International Multidisciplinary Research Foundation), IMFR Inventor Award 2020 , the Best Presentation Award at the 8th

International Conference on Engineering Mathematics and Physics ICEMP 2019 in Ningbo, China , Excellent award (Silver) of the virtual International, Invention, Innovation & Design Competition 2020 (3iDC2020) and Jury in the International Teaching and Learning Invention, Innovation Competition (iTaLiiC2023).

Analytical articles published in local and international media. Author for more than 60 Books , 8 preprint articles published in the World Health Organization (WHO) , 41 article published in the Europe PMC and 22 are full article.

Peter Chew also is CEO PCET, Ventures, Malaysia, PCET is a long research associate of IMRF (International Multidisciplinary Research Foundation), Institute of higher Education & Research with its HQ at India and Academic Chapters all over the world, PCET also Conference Partner in CoSMEd2021 by SEAMEO RECSAM.

Peter Chew as 2nd Plenary Speaker the 6th International Multidisciplinary Research Conference with a Mindanao Zonal Assembly on January 14, 2023, at the Immaculate Conception University, Bajada Campus, Davao City.

Keynote Speaker of the 8th International Conference on Computer Engineering and Mathematical Sciences (ICCEMS 2019) , the International Conference on Applications of Physics , Chemistry & Engineering Sciences, ICPCE 2020 , 2nd Global Summit on Public Health and Preventive Medicine (GSPHPM2023) June 19, 2023 and World BIOPOLYMERS & POLYMER CHEMISTRY CONGRESS" 10-11 July 2023 | Online by Drug Delivery,

Special Talk Speaker at the 2019 International Conference on Advances in Mathematics, Statistics and Computer Science, the 100th CONF of the IMRF,2019, Goa , India.

Invite Speaker of the 24th Asian Mathematical Technology Conference (ATCM 2019) Leshan China , the 5th(2020), 6th (2021) and 7th (2022) International Conference on Management, Engineering, Science, Social Sciences and Humanities by [Society For Research Development](SRD) and 12th International Conference on Engineering Mathematics and Physics (July 5-7, 2023 in Kuala Lumpur, Malaysia).

Peter Chew is also Program Chair for the 11th International Conference on Engineering Mathematics and Physics (ICEMP 2022, Saint-Étienne, France | July 7-9, 2022) and Program Chair for the 12th International Conference on Engineering Mathematics and Physics (ICEMP 2023, Kuala Lumpur, Malaysia | July 5-7, 2023).

For more information, please get it from this link Orcid: https://orcid.org/0000-0002-5935-3041.

PIONEERING TOMORROW'S AI SYSTEM THROUGH ELECTRICAL ENGINEERING AN EMPIRICAL STUDY OF THE PETER CHEW RULE FOR OVERCOMING ERROR IN CHAT GPT

TABLE OF CONTENTS

Pioneering Tomorrow's AI System Through Electrical Engineering. An Empirical Study Of The Peter Chew Rule For Overcoming Error In Chat GPT 9

1. Background. 13
   1.1 Chat GPT 13
   1.2 Introducing Chat GPT Plus 13
   1.3 Knowledge is power: why the future is not just about the tech. 14
   1.4 Electrical Engineering 15
   1.5 Electronics Tutorials.
       Resultant Value of VT 17
   1.6 Prove of Peter Chew Rule 18
   1.7 Memorization Techniques for Peter Chew Rule 19
   1.8 Peter Chew Triangle Diagram 21

2. Current Method and Peter Chew Rule Method 22

3. Application of Peter Chew Rule to Electrical Engineering 24

4. Pioneering Tomorrow's AI System Through Electrical Engineering 26

PIONEERING TOMORROW'S AI SYSTEM
THROUGH ELECTRICAL ENGINEERING
AN EMPIRICAL STUDY OF THE PETER CHEW RULE
FOR OVERCOMING ERROR IN CHAT GPT

TABLE OF CONTENTS

4.1 Electrical Engineering problems that can be solved directly by the Cosines Rule 26

4.2 Electrical Engineering problems that can be solved directly by the Sine Rule 29

4.3 Electrical Engineering problems that can not be solved directly by the Cosine and Sine Rule. 33

4.4 Electrical Engineering problems that can not be solved directly by the Cosine and Sine Rule. 41

5. Results 50

6. Conclusion 52

7. Discussion 54

8. Implications and Future Research 56

9. Acknowledgment 58

10. Reference 62

Pioneering Tomorrow's AI System Through Electrical Engineering. An Empirical Study Of The Peter Chew Rule For Overcoming Error In Chat GPT

Abstract:

Introduction:

This empirical study investigates the Peter Chew Rule for Overcoming Error In Chat GPT. – on enhancing Chat GPT's competence in effectively solving **Electrical Engineering** problem. The integration of Artificial Intelligence (AI) into Electrical Engineering problem -solving has paved the way for innovative approaches. This study aim to showcase the important of Peter Chew Rule For Overcoming Error In AI System like GPT Chat.

Evidence:

Drawing upon empirical evidence, this study presents a comprehensive exposition of ChatGPT's adept utilization of Peter Chew Rule correct solving Electrical Engineering problem that cannot be solved directly by the cosine and sine rules. In stark contradistinction, the Method adopted by

ChatGPT's can not correct solving **Electrical Engineering** problem that cannot be solved directly by the cosine and sine rules. This underscores the pivotal role endowed by the Peter Chew Rule in amplifying the solving **Electrical Engineering** problem proficiencies intrinsic to AI systems like Chat GPT.

Result :

The findings derived from this study unveil a compelling and notable demonstration of ChatGPT's adept utilization of the Peter Chew Rule. This Rule approach has yielded outcomes that are both substantial and convincing, particularly in the context of solving **Electrical Engineering** problem that cannot be solved directly by the cosine and sine rules.

This study's results provide compelling evidence of ChatGPT's adept use of the Peter Chew Rule, enabling correct solving **Electrical Engineering** problem that cannot be solved directly by the cosine and sine rules. In contrast, when ChatGPT using current approach, ChatGPT can not correct solving **Electrical Engineering** problem that cannot be solved directly by the cosine and sine rules. This performance disparity underscores the vital role of the Peter Chew Rule in enhancing

AI systems' solving <u>Electrical Engineering</u> problem abilities, highlighting the transformative potential of diverse methodologies in advancing AI's mathematical prowess.

Conclusion :

Pioneering Novel Maths Rule such as Peter Chew Rule for Solution of Triangle For Overcoming Errors in AI System like GPT Chat. This study underscores the importance of pioneering innovative Rule to overcome existing Errors in AI systems like ChatGPT, particularly in Solving Triangle Problem. The groundbreaking Peter Chew Rule for Solution of Triangle showcased here holds the promise of unleashing untapped potential, elevating AI systems to new levels of proficiency. Essentially, the Peter Chew Rule offers a strategic avenue for enhancing AI capabilities and pushing the boundaries of achievable accomplishments.

Discussion:

The outcomes derived from this study underscore the significant influence wielded by the method selection in augmenting the mathematical competencies of ChatGPT.

Particularly noteworthy is the application of the Peter Chew Rule, which surfaces as a compelling exemplar. This Rule serves as a overcomes current Errors on solving Electrical Engineering problem that cannot be solved directly by the cosine and sine rules in AI systems like ChatGPT.

Implications and Future Research:

These findings not only contribute to enhance AI's mathematical competencies but also emphasize the need for pioneering new Rules, Theorems, Methods or Formulas to further enhance AI systems like ChatGPT. Future research could explore the development of novel mathematical techniques tailored to AI systems, thus expanding their capabilities across diverse problem-solving domains. This can be effective in let Electrical Engineering student interest in using AI systems like ChatGPT while learning Electrical Engineering especially when analogous covid- 19 issues arise in the future.

Keyword: Peter Chew Rule for Solution of Triangle, AI System, ChatGPT, Electrical Engineering.

1. Background.

1.1 ChatGPT[1] :

Introducing ChatGPT.

We've trained a model called ChatGPT which interacts in a conversational way. The dialogue format makes it possible for ChatGPT to answer follow up questions, admit its mistakes, challenge incorrect premises, and reject inappropriate requests.

1.2 Introducing ChatGPT Plus

We're launching a pilot subscription plan for ChatGPT, a conversational AI that can chat with you, answer follow-up questions, and challenge incorrect assumptions. **Chat GPT**[2] knowledge is still limited to 2021 data, which means it can't answer current questions.

1.3 Knowledge is power: why the future is not just about the tech[3] .**If we are to rely on machine intelligence, we need to understand the two types of knowledge.**

Understanding knowledge means we can distinguish where we want machines to do the mundane work and where we want humans to perform intuitive tasks.Such an approach will be as beneficial for business as for education.

As virtual and physical worlds become increasingly interdependent, knowledge – and how we manage it – will become the secret ingredient to manage the situation. And thrive. Virtual technologies are swiftly becoming intertwined with our physical world, and companies need to adapt. But that doesn't simply mean replacing humans with robots or relying on artificial intelligence (AI) to make all of our decisions.

This is because technology, though powerful, is just part of the equation. In fact, human intelligence will be one of the most valuable assets in today's Fourth Industrial Revolution (FIR), and companies may flounder if they fail to strike the right balance of automated technology and human insights.

1.4 Electrical Engineering[4].

Lesson 5. PHASE RELATIONS AND VECTOR REPRESENTATION

Parallelogram method

This technique is used for addition of two phasors at a time. The two alternating quantities are denoted by phasor diagram. The two phasors are arranged as the adjacent sides of a parallelogram. The diagonal of the formal parallelogram gives the resultant value of the two phasors. The following diagram shows phasor diagram of a.c parallel circuit:

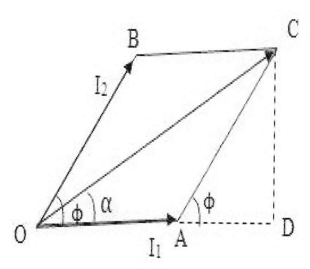

Fig 5.4 Phasor diagram of a.c parallel circuit

[A PHASOR DIAGRAM FOR AN AC CIRCUIT]

The two currents flowing in the circuit are given as:

$i_1 = I_{m1} \sin \omega t$

$i_2 = I_{m2} \sin(\omega t + \phi)$

i_r = resultant current

I_{m1} and I_{m2} are the maximum value of currents i_1 and i_2 respectively. Here i_1 is leading w.r.t i_2 or in other words i_2 is lagging w.r.t i_1. The phase difference between i_1 and i_2 is $\phi°$.

$$OC = \sqrt{(OD)^2 + (DC)^2}$$
$$= \sqrt{(OA + AD)^2 + (DC)^2}$$
$$= \sqrt{I_{m1}^2 + I_{m1}^2(\sin^2\phi + \cos^2\phi) + 2I_{m1}I_{m2}\cos\phi}$$
$$= \sqrt{I_{m1}^2 + I_{m2}^2 + 2I_{m1}I_{m2}\cos\phi}$$

$$\tan \alpha = \frac{y}{x} = \frac{CD}{OD}$$

Phase angle $\alpha = \tan^{-1}\frac{CD}{OD}$

$$= \tan^{-1}\frac{I_{m2}\sin\phi}{I_{m1} + I_{m2}\cos\phi}$$

The equation for instantaneous value of resultant current i_r is given as: $i_r = I_{mr} \sin(\omega t + \alpha)$

1.5 Electronics Tutorials. Resultant Value of VT [5].

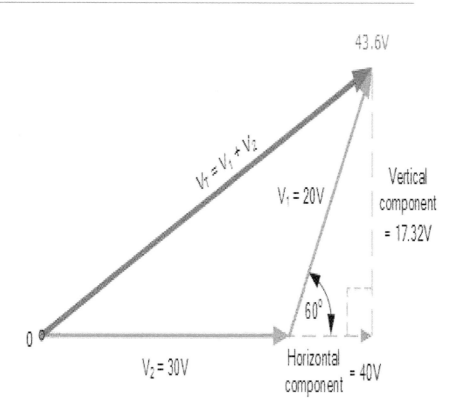

Phasor Subtraction of Phasor Diagrams

Phasor subtraction is very similar to the above rectangular method of addition, except this time the vector difference is the other diagonal of the parallelogram between the two voltages of V_1 and V_2 as shown.

1.6 Prove of Peter Chew Rule [6,7]

Fig. 1, $\tan \angle C = \dfrac{c \sin \angle B}{a - c \cos \angle B}$

$\cos \angle B = \dfrac{a_1}{c}$

$a_1 = c \cos \angle B \ldots\ldots i)$

$\sin \angle B = \dfrac{h}{c}$

$h = c \sin \angle B \ldots\ldots ii)$

$\tan \angle C = \dfrac{h}{a_2}$

$= \dfrac{h}{a - a_1}$

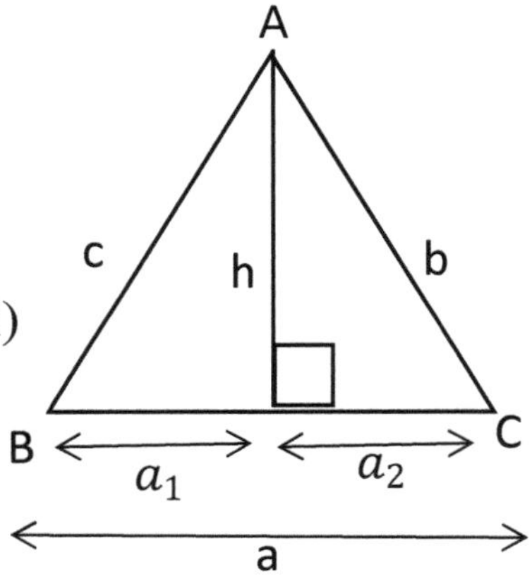

Fig. 1

From i) and ii), $\therefore \tan \angle C = \dfrac{c \sin \angle B}{a - c \cos \angle B}$

Mathematical Proof

Agata Stefanowicz from University of Birmingham that Mathematical mention proof is absolute, which means once a theorem is proved, it is proved forever. Until a proven though, the statement is never accepted as a true one. In my opinion, no one can claim that he created a rule unless he can prove it. Similarly, no one can conclude that someone is guilty unless he has evidence to prove it.

1.7 Memorization Techniques for Peter Chew Rule[8]

1. Write tan (the angle you want to find) on the left hand side, and the side opposite the angle on the right hand side for numerator and denominator, $tan \angle A = \dfrac{a}{a}$.

2. For the numerator, write sin (the angle given) and for the denominator, write cos (the angle given).

   i) If the angle is known as B, $tan \angle A = \dfrac{a \sin \angle B}{a \cos \angle B}$,

   ii) If the angle is known as C, $tan \angle A = \dfrac{a \sin \angle C}{a \cos \angle C}$.

3i) If the angle given is B, and side is a. For the denominator, will be $c - a \cos \angle B$, Memorization Techniques is denominator will involve a, b, c . So, remember the Peter Chew rule, Just as easy to remember a, b, c . We get Peter Chew Rule , $\tan \angle A = \dfrac{a \sin \angle B}{c - a \cos \angle B}$.

3ii) If the angle given is C, and side is a. For the denominator, b - a $\bm{cos \angle C}$, Memorization Techniques is denominator will involve a, b, c . So, remember the Peter Chew rule, Just as easy to remember a, b, c . We get Peter Chew Rule , $\bm{\tan \angle A} = \dfrac{a \sin \angle B}{b - a \cos \angle B}$.

By symmetry, 6 similar Peter Chew Rules[9]

$\tan \angle A = \dfrac{a \sin \angle B}{c - a \cos \angle B}$	$\tan \angle A = \dfrac{a \sin \angle C}{b - a \cos \angle C}$
$\tan \angle B = \dfrac{b \sin \angle A}{c - b \cos \angle A}$	$\tan \angle B = \dfrac{b \sin \angle C}{a - b \cos \angle C}$
$\tan \angle C = \dfrac{c \sin \angle A}{b - c \cos \angle A}$	$\tan \angle C = \dfrac{c \sin \angle B}{a - c \cos \angle B}$

1.8 Peter Chew Triangle Diagram[10]

Peter Chew Rule for Solution of Triangle enhances self-learning capabilities ; Applying Peter Chew Rule to Peter Chew Triangle Diagram can improve students' self-learning ability because Peter Chew Rules allows diagrams to guide students to simple solutions with only one rule.

Peter Chew Triangle Diagram

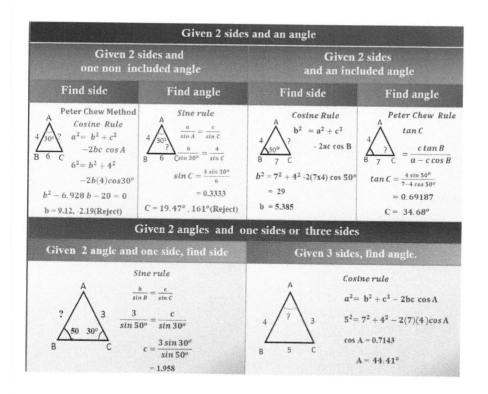

2. Current Method and Peter Chew Rule Method.

Example: Given ∠B = 35°, AB = 6 cm and BC = 3 cm. Find ∠C.

Solution:

Current Method 1: Cosine Rule plus Sine rule,

Cosine Rule, $b^2 = a^2 + c^2 - 2ac \cos\angle B$

$$b^2 = 3^2 + 6^2 - 2(3)(6) \cos \angle 35°$$

$$= 15.51$$

$$b = 3.938$$

Sine Rule, $\dfrac{b}{\sin\angle B} = \dfrac{c}{\sin\angle C}$

$$\dfrac{3.938}{\sin\angle 35°} = \dfrac{6}{\sin\angle C}$$

$$\sin \angle C = \dfrac{6 \sin 35°}{3.938}$$

$$= 0.8739$$

∠C = 119.0849°, 60.9151° (Reject)

Current Method 2: Cosine Rule plus Cosine rule,

Cosine Rule, $b^2 = a^2 + c^2 - 2ac \cos \angle B$

$b^2 = 3^2 + 6^2 - 2(3)(6) \cos \angle 35°$

$\quad = 15.51$

$b = 3.938$

Cosine Rule, $c^2 = a^2 + b^2 - 2ab \cos \angle C$

$6^2 = 3^2 + 3.938^2 - 2(3)(3.938) \cos \angle C$

$23.628 \cos \angle C = -11.49$

$\quad \cos \angle C = -0.4863$

$\quad \angle C = 119.0977°$

Peter Chew Rules . $\quad tan\angle C = \dfrac{c \sin \angle B}{a - c \cos \angle B}$

$\quad tan\angle C = \dfrac{6 \sin 35°}{3 - 6 \cos 35°}$

$\quad \quad = -1.797$

$\quad \angle C = 119.952°$

Note: The actual answer is 119.0926395°

3. Application of Peter Chew Rule to Electrical Engineering

Higher Engineering Mathematics.

Fifth Edition[11] (page 128).

Problem 26(page 128). Two voltage phasors are shown in Fig. 12.26. If $V_1 = 40V$ and $V_2 = 100V$ determine the angle the resultant makes with V_1.

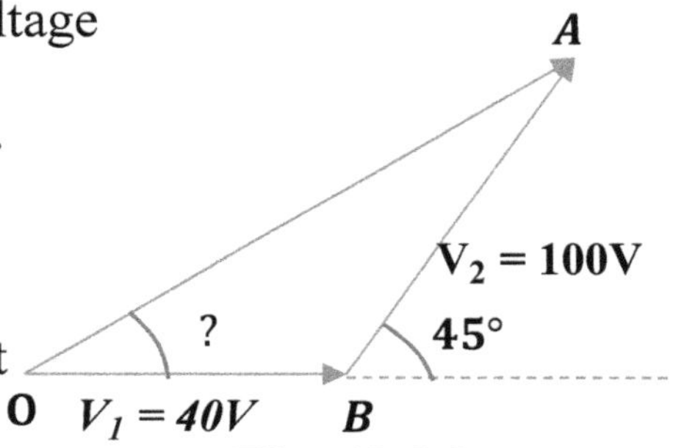

Fig. 12.26

Current Solution:

$\angle$ OBA $= 180° - 45° = 135°$

Applying the cosine rule,

$OA^2 = V_1^2 + V_2^2 - 2\,V_1\,V_2 \cos \angle OBA$

$\qquad = 40^2 + 100^2 - 2\,(40)(100) \cos 135°$

$\qquad = 17257$

The resultant, OA = $\sqrt{17257}$ = 131.4V

Applying the sine rule: $\dfrac{131.4}{\sin 135°} = \dfrac{100}{\sin \angle AOB}$

$$\sin \angle AOB = \dfrac{100 \sin 135°}{131.4}$$

$$= 0.5381$$

∠ AOB = **32°33`, 147°27`** (which is impossible in this case)

Hence the angle the resultant makes with V_1 is 32°33`

Peter Chew Rule For Solution Of Triangle

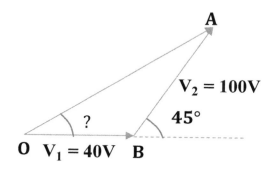

$$\tan \angle AOB = \dfrac{100 \sin 135°}{40 - 100 \cos 135°}$$

$$= 0.6387$$

∠ AOB = 32°34`

Hence the angle the resultant makes with V_1 is 32°34`

4. Pioneering Tomorrow's AI System Through Electrical Engineering

4.1 Electrical Engineering problems that can be solved directly by the Cosines Rule

Example 4.1 A phasor diagram for an AC circuit can be show in triangle ABC.

If $V_1 = AB = 20$ V , $V_2 = BC = 80$ V and angle ABC= **120°**. Determine the resultant(AC).

Current Solution

$\angle OBA = 180° - 60° = 120°$

Applying the cosine rule,

$$AC^2 = V_1^2 + V_2^2 - 2\, V_1\, V_2 \cos \angle ABC$$

$$= 20^2 + 80^2 - 2(20)(80) \cos 120°$$

$$= 8\,400$$

The resultant, $AC = \sqrt{8\,400} = 91.65$ V

Chat **GPT**[12] can correctly solve **Electrical Engineering** problems that can be solved directly by the Cosines Rule.

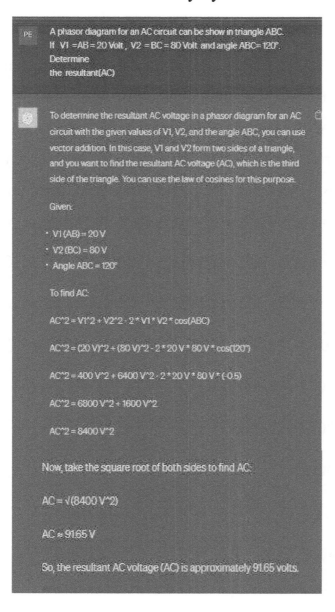

Full text: A phasor diagram for an AC circuit can be show in triangle ABC. If V1 =AB = 20 Volt , V2 = BC = 80 Volt and angle ABC= 120°. Determine the resultant(AC)

ChatGPT: To determine the resultant AC voltage in a phasor diagram for an AC circuit with the given values of V1, V2, and the angle ABC, you can use vector addition. In this case, V1 and V2 form two sides of a triangle, and you want to find the resultant AC voltage (AC), which is the third side of the triangle. You can use the law of cosines for this purpose.

Given: V1 (AB) = 20 V. V2 (BC) = 80 V. Angle ABC = 120°

To find AC: AC^2 = V1^2 + V2^2 - 2 * V1 * V2 * cos(ABC)

AC^2 = (20 V)^2 + (80 V)^2 - 2 * 20 V * 80 V * cos(120°)

AC^2 = 400 V^2 + 6400 V^2 - 2 * 20 V * 80 V * (-0.5)

AC^2 = 6800 V^2 + 1600 V^2

AC^2 = 8400 V^2

Now, take the square root of both sides to find AC:

AC = √(8400 V^2)

AC ≈ 91.65 V

So, the resultant AC voltage (AC) is approximately 91.65 volts.

4.2 Electrical Engineering problems that can be solved directly by the Sine Rule.

Chat GPT can correctly solve Electrical Engineering [Jib Crane] problems that can be solved directly by the Sine Rule

Example 4.2 A phasor diagram for an AC circuit can be show in triangle ABC.

If $V_2 = BC = 80$ V , the resultant(AC) = 91.65 V and angle ABC= 120°.

Determine the angle CAB [angle of the resultant(AC) makes with V_1 (AB)].

Sine rule:

$$\frac{91.65}{\sin 120°} = \frac{80}{\sin \angle CAB}$$

$$\sin \angle CAB = \frac{80 \sin 120°}{91.65}$$

$$= 0.75594$$

$\angle$ CAB=**49.108°**,**130.9°** (which is impossible in this case)

Hence the angle the resultant makes with V_1 is **49.108°** .

Chat **GPT**[12] can correctly solve **Electrical Engineering** problems that can be solved directly by the Sine Rule.

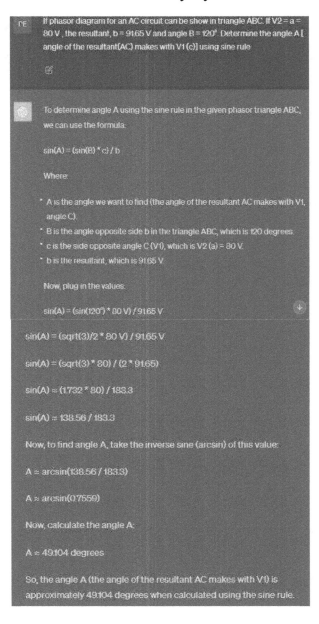

Full Text: If phasor diagram for an AC circuit can be show in triangle ABC. If V2 = a = 80 V, the resultant, b = 91.65 V and angle B = 120°. Determine the angle A [angle of the resultant(AC) makes with V1 (c)] using sine rule

ChatGPT

To determine angle A using the sine rule in the given phasor triangle ABC, we can use the formula:

sin(A) = (sin(B) * c) / b

Where:

A is the angle we want to find (the angle of the resultant AC makes with V1, angle C).

B is the angle opposite side b in the triangle ABC, which is 120 degrees.

c is the side opposite angle C (V1), which is V2 (a) = 80 V.

b is the resultant, which is 91.65 V.

Now, plug in the values:

sin(A) = (sin(120°) * 80 V) / 91.65 V

$\sin(A) = (\text{sqrt}(3)/2 * 80 \text{ V}) / 91.65 \text{ V}$

$\sin(A) = (\text{sqrt}(3) * 80) / (2 * 91.65)$

$\sin(A) \approx (1.732 * 80) / 183.3$

$\sin(A) \approx 138.56 / 183.3$

Now, to find angle A, take the inverse sine (arcsin) of this value:

$A \approx \arcsin(138.56 / 183.3)$

$A \approx \arcsin(0.7559)$

Now, calculate the angle A:

$A \approx 49.104$ degrees

So, the angle A (the angle of the resultant AC makes with V1) is approximately 49.104 degrees when calculated using the sine rule.

4.3 Electrical Engineering problems that can not be solved directly by the Cosine and Sine Rule.

Example 4.3 : A phasor diagram for an AC circuit can be show in triangle ABC.

If $V_1 = AB = 20$ V, $V_2 = BC = 80$ V and angle ABC = 120°. Determine the angle of the resultant (AC) makes with V_1 (AB).

Current Solution

$\angle OBA = 180° - 60° = 120°$

Applying the cosine rule,

$AC^2 = V_1^2 + V_2^2 - 2 V_1 V_2 \cos \angle ABC$

$\quad = 20^2 + 80^2 - 2(20)(80) \cos 120°$

$\quad = 8\,400$

The resultant, $AC = \sqrt{8\,400} = 91.65$ V

Applying the sine rule: $\dfrac{91.65}{\sin 120°} = \dfrac{80}{\sin \angle CAB}$

$$\sin \angle \text{CAB} = \frac{80 \sin 120°}{91.65}$$

$$= 0.75594$$

∠ CAB=**49.108°, 130.9°** (which is impossible in this case)

Hence the angle the resultant makes with V_1 is 49.108° .

Peter Chew Rule For Solution Of Triangle

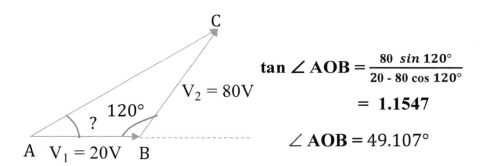

$$\tan \angle \text{AOB} = \frac{80 \sin 120°}{20 - 80 \cos 120°}$$

$$= 1.1547$$

$$\angle \text{AOB} = 49.107°$$

Hence the angle the resultant makes with V_1 is 49.107° .

Actual answer: 49..10660535....

Note: Besides being simple, Peter Chew's rule is also more accurate than current methods, because the main advantage of Peter Chew's rule is that there is no square root step[cosine rule]. Usually the square root value is only an approximation

4.3.1 Chat GPT cannot correctly solve Electrical Engineering problems that can not be solved directly by the Cosine and Sine Rule .

Example 4.3 : If phasor diagram for an AC circuit can be show in triangle ABC. If $V1 = c = 20$ V and $V2 = a = 80$ V and angle B = 120°. Determine the angle A [angle of the resultant(b) makes with V1 (c)] .

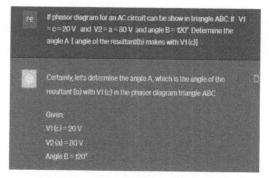

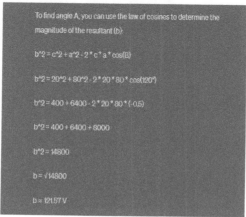

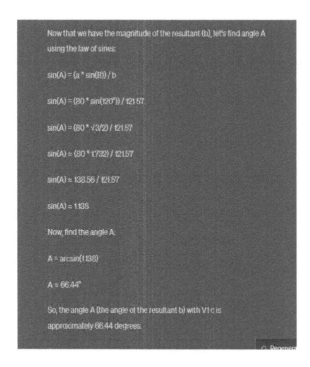

NOTE: Chat GPT cannot correctly solve the above Electrical Engineering problem [which cannot be solved directly by cosine and sine rules] probably because the solution is complex and involves many steps.

Therefore, we need to discover new simple rules to solve the above problems. This is one of the goals of the simple rule, Peter Chew's rule.

Full text. **Wrong step[Bold text]**

If phasor diagram for an AC circuit can be show in triangle ABC. If V1 = c = 20 V and V2 = a = 80 V and angle B = 120°. Determine the angle A [angle of the resultant(b) makes with V1 (c)] .

ChatGPT:

Certainly, let's determine the angle A, which is the angle of the resultant (b) with V1 (c) in the phasor diagram triangle ABC. Given: V1 (c) = 20 V V2 (a) = 80 V Angle B = 120°

To find angle A, you can use the law of cosines to determine the magnitude of the resultant (b): $b^2 = c^2 + a^2 - 2 * c * a * \cos(B)$

$b^2 = 20^2 + 80^2 - 2 * 20 * 80 * \cos(120°)$

$b^2 = 400 + 6400 - 2 * 20 * 80 * (-0.5)$

$b^2 = 400 + 6400 + 8000$

$b^2 = 14800$

$b = \sqrt{14800}$

$b \approx 121.57$ V

Now that we have the magnitude of the resultant (b), let's find angle A using the law of sines:

$\sin(A) = (a * \sin(B)) / b$

$\sin(A) = (80 * \sin(120°)) / 121.57$

$\sin(A) = (80 * \sqrt{3}/2) / 121.57$

$\sin(A) \approx (80 * 1.732) / 121.57$

$\sin(A) \approx 138.56 / 121.57$

$\sin(A) \approx 1.138$

Now, find the angle A:

$A = \arcsin(1.138)$

$A \approx 66.44°$

So, the angle A (the angle of the resultant b) with V1 c is approximately 66.44 degrees.

4.3.2 Chat GPT uses Peter Chew's rules to correctly solve Electrical Engineering problems that cannot be solved directly by cosine and sine rules.

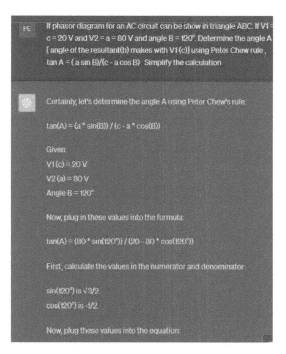

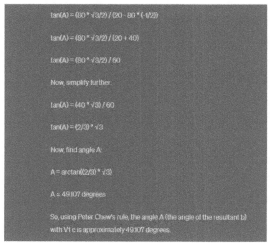

Note: Chat GPT can use Peter Chew's rules to correctly solve Electrical Engineering problems that cannot be solved directly by cosine and sine rules because the solution becomes simple and involves fewer steps.

4.4 Electrical Engineering problems that can not be solved directly by the Cosine and Sine Rule.

Example 4.4 If phasor diagram for an AC circuit can be show in triangle ABC. If V1 = c = 30 V and V2 = a = 50 V and angle B = 100°. Determine the angle A [angle of the resultant(b) makes with V1 (c)] .

Current Method 1: Cosine Rule plus Sine rule,

Cosine Rule, $b^2 = a^2 + c^2 - 2ac \cos\angle B$

$$b^2 = 50^2 + 30^2 - 2(50)(30) \cos 100°$$

$$= 3921$$

$$b = 62.62 \text{ Volt}$$

Sine Rule, $\dfrac{b}{\sin\angle B} = \dfrac{a}{\sin\angle A}$

$$\dfrac{62.62}{\sin 100°} = \dfrac{50}{\sin\angle A}$$

$$\sin\angle A = \dfrac{50 \sin 100°}{62.62}$$

$$= 0.7863$$

$\angle A = 51.84°, 128.16°$ (Reject)

Current Method 2: Cosine Rule plus Cosine rule,

Cosine Rule, $b^2 = a^2 + c^2 - 2ac \cos \angle B$

$b^2 = 50^2 + 30^2 - 2(50)(30) \cos 100°$

$ = 3921$

$b = 62.62$ Volt

Cosine Rule, $a^2 = b^2 + c^2 - 2bc \cos A$

$50^2 = 62.62^2 + 30^2 - 2(62.62)(30) \cos A$

$3757 \cos A = 2321$

$ \cos A = 0.6178$

$ \angle A = 51.84°$

Peter Chew Rules Method.

$$\tan A = \frac{50 \sin 100°}{30 - 50 \cos 100°}$$

$$= 1.273$$

$$\angle A = 51.85°$$

Drift angle, $\angle A$ is $51.85°$

Note: Actual Answer is 51.84742646°

4.4.1 Chat GPT cannot correctly solve **Aerospace Engineering** problems that can not be solved directly by the Cosine and Sine Rule.

Example 4.4 If phasor diagram for an AC circuit can be show in triangle ABC. If V1 = c = 30 V and V2 = a = 50 V and angle B = 100°. Determine the angle A [angle of the resultant(b) makes with V1 (c)].

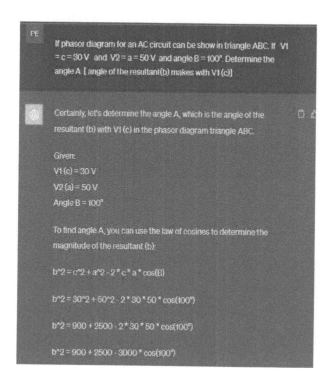

Now, calculate b:

$b = \sqrt{(900 + 2500 - 3000 * \cos(100°))}$

$b \approx \sqrt{(3400 - 3000 * \cos(100°))}$

$b \approx \sqrt{(3400 - 3000 * (-0.1736))}$

$b \approx \sqrt{(3400 + 520.8)}$

$b \approx \sqrt{3920.8}$

$b \approx 62.59\ V$

Now that we have the magnitude of the resultant (b), let's find angle A using the law of sines:

$\sin(A) = (a * \sin(B)) / b$

$\sin(A) = (50 * \sin(100°)) / 62.59$

$\sin(A) \approx (50 * 0.1736) / 62.59$

$\sin(A) \approx 0.0868$

Now, find the angle A:

$A = \arcsin(0.0868)$

$A \approx 4.98°$

So, the angle A (the angle of the resultant b) with V1 c is approximately 4.98 degrees.

NOTE: Chat GPT cannot correctly solve the above Electrical e engineering problem [which cannot be solved directly by cosine and sine rules] probably because the solution is complex and involves many steps. Therefore, we need to discover new simple rules to solve the above problems. This is one of the goals of the simple rule, Peter Chew's rule.

Full text: Wrong step [Red Bold text]

If phasor diagram for an AC circuit can be show in triangle ABC. If V1 = c = 30 V and V2 = a = 50 V and angle B = 100°. Determine the angle A [angle of the resultant(b) makes with V1 (c)] .

ChatGPT

Certainly, let's determine the angle A, which is the angle of the resultant (b) with V1 (c) in the phasor diagram triangle ABC.

Given: V1 (c) = 30 V V2 (a) = 50 V Angle B = 100°

To find angle A, you can use the law of cosines to determine the magnitude of the resultant (b):

b^2 = c^2 + a^2 - 2 * c * a * cos(B)

b^2 = 30^2 + 50^2 - 2 * 30 * 50 * cos(100°)

b^2 = 900 + 2500 - 2 * 30 * 50 * cos(100°)

b^2 = 900 + 2500 - 3000 * cos(100°)

Now, calculate b:

b = √(900 + 2500 - 3000 * cos(100°))

$b \approx \sqrt{(3400 - 3000 * \cos(100°))}$

$b \approx \sqrt{(3400 - 3000 * (-0.1736))}$

$b \approx \sqrt{(3400 + 520.8)}$

$b \approx \sqrt{3920.8}$

$b \approx 62.59$ V

Now that we have the magnitude of the resultant (b), let's find angle A using the law of sines:

$\sin(A) = (a * \sin(B)) / b$

$\sin(A) = (50 * \sin(100°)) / 62.59$

$\sin(A) \approx (50 * 0.1736) / 62.59$ [$\sin(100°) = 0.9848$]

$\sin(A) \approx 0.0868$

Now, find the angle A:

$A = \arcsin(0.0868)$

$A \approx 4.98°$

So, the angle A (the angle of the resultant b) with V1 c is approximately 4.98 degrees.

47

4.4.2 Chat GPT uses Peter Chew's rules to correctly solve Electrical engineering problems that cannot be solved directly by cosine and sine rules

Example 4.4 If phasor diagram for an AC circuit can be show in triangle ABC. If $V1 = c = 30$ V and $V2 = a = 50$ V and angle $B = 100°$. Determine the angle A [angle of the resultant(b) makes with V1 (c)] using Peter Chew rule , $\tan A = (a \sin B)/(c - a \cos B)$. Simplify the calculation.

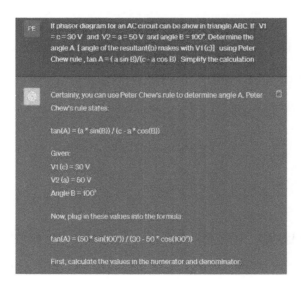

sin(100°) is approximately 0.9848.
cos(100°) is approximately -0.1736.

Now, plug these values into the equation:

tan(A) = (50 * 0.9848) / (30 - 50 * (-0.1736))

tan(A) = (49.24) / (30 + 8.68)

tan(A) = 49.24 / 38.68

Now, calculate tan(A):

tan(A) ≈ 1.27

Now, find angle A:

A = arctan(1.27)

A ≈ 51.64°

So, using Peter Chew's rule, the angle A (the angle of the resultant b) with V1 c is approximately 51.64 degrees.

Note: Chat GPT can use Peter Chew's rules to correctly solve Electrical Engineering problems that cannot be solved directly by cosine and sine rules because the solution becomes simple and involves fewer steps.

Full text:

If phasor diagram for an AC circuit can be show in triangle ABC. If V1 = c = 30 V and V2 = a = 50 V and angle B = 100°. Determine the angle A [angle of the resultant(b) makes with V1 (c)] using Peter Chew rule , tan A = (a sin B)/(c - a cos B) . Simplify the calculation

ChatGPT

Certainly, you can use Peter Chew's rule to determine angle A.

Peter Chew's rule states:

$\tan(A) = (a * \sin(B)) / (c - a * \cos(B))$

Given: V1 (c) = 30 V V2 (a) = 50 V Angle B = 100°

Now, plug in these values into the formula:

$\tan(A) = (50 * \sin(100°)) / (30 - 50 * \cos(100°))$

First, calculate the values in the numerator and denominator:

sin(100°) is approximately 0.9848. cos(100°) is approximately -0.1736.

Now, plug these values into the equation:

$\tan(A) = (50 * 0.9848) / (30 - 50 * (-0.1736))$

$\tan(A) \approx (49.24) / (30 + 8.68)$

$\tan(A) \approx 49.24 / 38.68$

Now, calculate tan(A):

$\tan(A) \approx 1.27$

Now, find angle A:

$A = \arctan(1.27)$

$A \approx 51.64°$

So, using Peter Chew's rule, the angle A (the angle of the resultant b) with V1 c is approximately 51.64 degrees.

5. Results

The findings derived from this study unveil a compelling and notable demonstration of Chat GPT's adept utilization of the Peter Chew Rule. This rule approach has yielded outcomes that are both substantial and convincing, particularly in the context of solving Electrical Engineering problem that cannot be solved directly by the cosine and sine rules.

Contrastingly, when employing the conventional method, Chat GPT can not correct solving **Electrical Engineering** problem that cannot be solved directly by the cosine and sine rules **probably because the solution is complex and involves many steps.**

This discrepancy in performance between the two approaches serves to highlight a significant distinction, underscoring the pivotal role played by the Peter Chew Rule in elevating the AI system's solving **Electrical Engineering** problem proficiency to new heights.

The implications of this research findings are profound and resonant. This study emphasize not only the immediate advancements achieved through the integration of the Peter Chew Rule but also the overarching potential it holds for the broader landscape of AI and **Electrical Engineering** problem - solving solution.

This stark contrast in outcomes serves as a testament to the transformative power that diverse methodologies can exert in shaping the capabilities of AI systems and unlocking novel avenues of solving **Electrical Engineering** problem solution exploration and comprehension.

6. Conclusion

Pioneering Novel Math's Rule such as Peter Chew Rule for Solution of Triangle For Overcoming Errors [Solving Electrical Engineering problem] in AI System like GPT Chat.

The groundbreaking Peter Chew Rule has emerged as a transformative force in the realm of mathematics, particularly within the context of overcoming the persistent issue of giving correct answer for **solving Electrical Engineering problem** in AI systems like GPT Chat.

This study stands as a resounding testament to the imperative nature of pioneering inventive Rule to address such challenges, as exemplified by the Peter Chew Rule.

By delving into uncharted territories and devising innovative solutions, this study sheds light on the remarkable potential that lies in adopting novel approaches. The Peter Chew Rule, specifically designed for giving correct answer , direct and simple **solving Electrical Engineering problem**, emerges as a beacon of promise, paving the way for AI systems, like ChatGPT, to transcend their Error. This Rule innovation offers a compelling path forward, promising to elevate the proficiency of AI systems to unprecedented levels.

In essence, the Peter Chew Rule acts as a strategic gateway to expanding the horizons of AI capabilities. It embodies a fresh perspective that not only addresses Error of AI System like Chat GPT but also charts a course toward pushing the boundaries of what AI can achieve.

This study serves as a reminder of the inherent power within inventive Rule, illustrating their role in shaping the trajectory of AI advancement. As AI systems continue to evolve , embracing pioneering Rule becomes not just an option, but a necessity for unlocking the full potential of artificial intelligence.

7. Discussion

The insights garnered from this study distinctly highlight the substantial role that the selection of methodology plays in elevating the mathematical capabilities of ChatGPT. Among the various approaches explored, the application of the Peter Chew Rule emerges as a particularly noteworthy and impactful example.

The Peter Chew Rule stands as a testament to innovation in overcoming challenges inherent in AI systems like ChatGPT. Specifically, it addresses the AI system like Chat GPT persisting issue of can not giving correct answer for solving **Electrical Engineering** problem that cannot be solved directly by the cosine and sine rules.

By seamlessly integrating this Rule, ChatGPT is empowered to transcend its previous Error and direct solving **Electrical Engineering** problem that cannot be solved directly by the cosine and sine rules.

Peter Chew Rule embodies a profound advancement in the AI's ability to tackle complex mathematical scenarios. Its integration signifies a pivotal turning point, underscoring the significance of innovative methodologies in reshaping the landscape of AI solutions.

In essence, the utilization of the Peter Chew Rule marks a decisive step towards refining ChatGPT's mathematical prowess. By effectively bridging the gap between existing capabilities and the pursuit of more comprehensive solutions, this Rule symbolizes the potential for exponential growth in AI's mathematical competencies.

As AI systems continue to evolve, methodologies like the Peter Chew Rule serve as guiding beacons, illuminating the path towards enhanced proficiency and groundbreaking achievements.

8. Implications and Future Research:

The discoveries unveiled through this findings hold the potential to revolutionize the mathematical aptitude of AI, while also underscoring the imperative for the continuous innovation of new Rule, Theorem, Methods or formulas to propel AI systems like ChatGPT to even greater heights.

This study serves as a clarion call for future research endeavors to delve into the creation of innovative mathematical techniques, meticulously tailored to the needs of AI systems.

This avenue of exploration promises to usher in a new era of AI capabilities, expanding their influence across a diverse spectrum of problem-solving domains. The pursuit of novel methodologies could yield solutions that resonate deeply with students, thereby enhancing their engagement with AI systems like ChatGPT—particularly relevant in the context of unforeseen challenges like those posed by events akin to the COVID-19 pandemic.

As we navigate the future, the symbiotic relationship between AI and mathematics is poised to thrive through inventive strategies. By developing methodologies attuned to AI systems, we unlock the potential for heightened student interest in leveraging tools like ChatGPT for learning solving Electrical Engineering problem.

This synergy could offer a valuable resource to mitigate learning disruptions during periods of crisis, fostering a dynamic and adaptable learning environment for students even when confronted with unprecedented circumstances.

9. Acknowledgement:

9.1 Peter Chew Rule has passed peer review by The International Conference on Engineering Mathematics and Physics, ICEMP 2019. Peter Chew Rule collected in Journal of Physics: Conference Series (*J. Phys.: Conf. Ser.* 1411 012009DOI 10.1088/1742-6596/1411/1/012009). Conference Proceedings Citation Index Scopus, Ei Compendex, etc.

Journal of Physics: Conference Series

PAPER • OPEN ACCESS

Peter Chew rule for solution of triangle

Peter Chew[1]
Published under licence by IOP Publishing Ltd
Journal of Physics: Conference Series, Volume 1411, 2019 the 8th International Conference on Engineering, Mathematics and Physics 1-3 July 2019, Ningbo, China
Citation Peter Chew 2019 *J. Phys.: Conf. Ser.* 1411 012009
DOI 10.1088/1742-6596/1411/1/012009

Certificate for Oral Presentation EMP

This Certificate is Awarded to

Peter Chew-HP1-3001

Paper Title:

Peter Chew Rule for Solution Of Triangle

For her/his attendance and delivery of an oral presentation in the 8th International Conference on Engineering Mathematics and Physics (ICEMP 2019) held in Ningbo, China on July 1-3, 2019.

Conference Committee
ICEMP 2019

9.2 Best Presentation Award for Pete Chew Rule at the 8th International Conference on Engineering Mathematics and Physics ICEMP 2019 in Ningbo, China. http://www.icemp.org/history.html .

9.3 Peter Chew Rule and Peter Chew Method are simple solution in Peter Chew Triangle Diagram and Peter Chew Triangle Diagram has passed peer review by The 12th International Conference on Engineering Mathematics and Physics, ICEMP 2023.

9.4 Peter Chew Triangle Diagram(preprint) is share at World Health Organization because the purpose of Peter Chew Triangle Diagram is to help teaching mathematics more easily , especially when similar covid-19 problems arise in the future. https://pesquisa.bvsalud.org/global-literature-on-novel-coronavirus-2019-ncov/resource/en/ppzbmed-10.20944.preprints202106.0221.v1

10. Reference:

1. Introducing ChatGPT. OpenAI . https://openai.com/blog/chatgpt

2. Matt G. Southern OpenAI's ChatGPT Update Brings Improved Accuracy. Search Engine Journal. January 10, 2023. https://www.searchenginejournal.com/openai-chatgpt-update/476116/#close

3. *James Lin* ,Knowledge is power: why the future is not just about the tech . World Economy Forum Jan 25, 2021 https://www.weforum.org/agenda/2021/01/knowledge-is-power-why-the-future-is-not-just-about-the-tech/

4. Lesson 5. PHASE RELATIONS AND VECTOR REPRESENTATION Electrical Engineering.

http://ecoursesonline.iasri.res.in/mod/resource/view.php?id=93227

5. Phasor Diagrams and Phasor Algebra. Electronics Tutorials

https://www.electronics-tutorials.ws/accircuits/phasors.html

6. Agata Stefanowicz, Joe Kyle, Michael Grove. Proofs and Mathematical Reasoning. University of Birmingham. September 2014

7. **Dr. Yibiao Pan. Mathematical Proofs and Their Importance. December 5, 2017 .**

8. Chew, Peter, Memorization Techniques for Peter Chew Rule (March 5, 2021). SSRN: https://ssrn.com/abstract=3798502 or http://dx.doi.org/10.2139/ssrn.3798502

9. Chew, Peter, Peter Chew Rule for Solution of Triangle (2019). 2019 the 8th International Conference on Engineering Mathematics and Physics, Journal of Physics: Conference Series 1411 (2019) 012009, IOP Publishing, doi:10.1088/1742-

6596/1411/1/012009, Available at SSRN: https://ssrn.com/abstract=3843433

10. Chew, P. Peter Chew Triangle Diagram and Application. *Preprints* 2021, 2021060221. https://doi.org/10.20944/preprints202106.0221.v2

11. John Bird. Higher Engineering Mathematics Fifth Edition. https://www.academia.edu/10275047/HIGHER_ENGINEERING_MATHEMATICS_In_memory_of_Elizabeth_Higher_Engineering_Mathematics_Fifth_Edition.

12. Chat *GPT* . https://chat.openai.com/?model=text-davinci-002-render-sha .

Milton Keynes UK
Ingram Content Group UK Ltd.
UKHW020727081123
432193UK00018B/716